GUIDELINES FOR PREDICTING DIETARY INTAKE OF PESTICIDE RESIDUES

Prepared by the
Joint UNEP/FAO/WHO
Food Contamination Monitoring Programme
in collaboration with the
Codex Committee on Pesticide Residues

World Health Organization
Geneva
1989

ISBN 92 4 154250 0

Printed in Switzerland

88/7900 - IAM - 5000

CONTENTS

PREFACE

The Codex Committee on Pesticide Residues (CCPR) is an inter-governmental body that advises the Codex Alimentarius Commission on all matters relating to pesticide residues. Its primary objective is to develop Codex maximum residue limits (MRLs), in order to facilitate international trade, while protecting the health of the consumer. Public health considerations are taken into account by establishing the MRLs at levels not higher than those resulting from use of the pesticide in accordance with good agricultural practice.

From time to time the question has been raised at the CCPR of whether acceptance of Codex MRLs could result in a situation in which the acceptable daily intake (ADI) of a pesticide would be exceeded. A definitive answer to this question can only be obtained by means of dietary intake studies (*1*). In cases where such studies are not feasible or where the pesticide has not long been in use, it is necessary to predict the pesticide residue intake on the basis of the available data.

The present guidelines describe procedures for predicting the dietary intake of pesticide residues, and are intended to assist national authorities in their considerations regarding the acceptability of Codex MRLs.

The initial draft of the guidelines was prepared by Dr R. D. Schmitt, Office of Pesticide Programs, Environmental Protection Agency, Washington, DC, USA. They were further developed by an FAO/WHO Consultation which met in Geneva on 5-8 October 1987 under the auspices of the Joint UNEP/FAO/ WHO Food Contamination Monitoring Programme (or GEMS/Food) in collaboration with the CCPR. GEMS/Food forms part of the Global Environment Monitoring System established by the United Nations Environment Programme. Dr Schmitt's continued help throughout all phases of the preparation of this publication is greatly appreciated.

Relevant authorities are invited to consider the basic approaches described here, which are designed to provide reasonable assurance that use of Codex MRLs will not result in a dietary intake of a pesticide that exceeds its ADI.

1. ACCEPTABLE DAILY INTAKE AND MAXIMUM RESIDUE LIMITS

The acceptable daily intake (ADI) of a pesticide is established by the Joint FAO/WHO Meeting on Pesticide Residues (JMPR), on the basis of a complete review of the available data (biochemical, metabolic, pharmacological, toxicological, etc.) from a wide range of experimental animal studies and observations in humans. The no-observed-adverse-effect level (NOAEL) for the most sensitive toxicological parameter, normally in the most sensitive species of experimental animal, is used as the starting-point. A safety factor that takes into consideration the type of effect, the severity or reversibility of the effect, and the problems of inter- and intra-species variability is applied to the NOAEL to determine the ADI for humans. Pertinent human data may outweigh experimental animal data in the estimation of the ADI for man.

Pesticide residue levels that would be expected from the use of good agricultural practice are estimated from globally generated data, and are likely to change as this practice is modified. The maximum residue limits (MRLs) recommended by the JMPR, on which the Codex MRLs are usually based, reflect the considered decision of the experts present at the meeting after examination of all pertinent data.

Neither the ADI nor the MRL is immutable. Both are determined according to the best judgement of a group of internationally recognized experts on the data available to them at the time of the evaluation. Summaries of these data are published in the JMPR Evaluations (6). However, as new data become available, the ADI or MRL may be revised.

From time to time, concern has been expressed over the possibility of adverse health effects arising from exposure to residues of more than one pesticide in food. This matter was considered by the 1981 JMPR (2), which concluded that, with the levels of pesticide residue intake found at that time, there was no need to alter the general approach for estimating ADIs. By extension of this conclusion, the approaches recommended here for the assessment of pesticide residue intake are also appropriate for the assessment of the concurrent intake of residues of more than one pesticide.

2. PREDICTING THE DIETARY INTAKE OF PESTICIDE RESIDUES

General considerations

In order to reach a conclusion as to the acceptability of an MRL from a public health point of view, it is necessary to predict the dietary intake of pesticide residues resulting from application of the MRL, and to compare this prediction with the ADI. The dietary intake of any particular pesticide residue in a given food is obtained by multiplying the residue level in the food by the amount of that food consumed. Total intake of the pesticide residue is then obtained by summing the intakes from all commodities containing the residue concerned.

Indices of residue level

Several indices of residue level can be used to predict pesticide residue intake. The MRL is one such index and represents the maximum residue level that is expected to occur in a commodity following the application of a pesticide according to good agricultural practice. Factors that may be taken into consideration when choosing an index to be used in predicting pesticide residue intake include the residue levels found in practice, their distribution in the commodity, and the effect on residues of the various processes used in the preparation of food.

It should be appreciated that the use of the MRL in the prediction of pesticide residue intake will lead to an overestimation of actual pesticide residue intake (3).

The prediction of intake of a particular pesticide residue should include all commodities for which MRLs have been established, unless the value has been estimated to be at, or about, the limit of determination.

Indices of food consumption

There are several possible indices of food consumption, a commonly used index being the average daily consumption. Others include average portion sizes, percentile consumption values, and the average consumption by people who actually eat the commodity. In predicting pesticide residue intake an effort should be made to reflect long-term food consumption habits and not day-to-day variations, in order to permit a valid comparison with the ADI, which is based on acceptable intake over a lifetime. Thus, it is recommended that average daily food consumption values be used in predicting pesticide residue intake for comparison with the ADI.

Food consumption patterns vary considerably from country to country and even within a country; thus, to a large extent, individual countries will have to estimate their own consumption pattern. However, for the purpose of predicting pesticide residue intake at the international level, the use of average food consumption data given in FAO Food Balance Sheets is recommended (4). Although the food consumption data derived from such sheets are subject to many uncertainties and limitations, they represent the best available source for international comparison and provide an approximate picture of the overall food situation in countries. Such an approximation of the overall patterns of food consumption is adequate for predicting pesticide residue intake, given the associated uncertainties in all of the components involved.

In order to predict pesticide residue intake at the international level, hypothetical diets will need to be developed for a number of dietary patterns that are representative of various regions of the world ("cultural" diets). As a first approximation, a hypothetical global diet consisting of the highest average value of food consumption for each "cultural" diet may suffice. Selection of this value for individual commodities from each "cultural" diet will, however, result in an unrealistic total food consumption. For the prediction of pesticide residue intake, these values should be normalized to a total daily consumption of 1.5 kg of solid food, i.e., excluding the liquid content of juices or milk.

For more realistic predictions, the "cultural" diets should be used individually. This would make it possible to predict a range of potential intakes.

For predictions of pesticide residue intake carried out at the national level, the best available food consumption data should be used. Countries should be cautious in the use of food consumption values other than average values, if such use results in a hypothetical level of consumption that would not be attained in practice. In carrying out predictions of pesticide residue intake for identifiable subgroups, e.g., vegetarians, it would be appropriate to use relevant average food consumption data for such subgroups.

Assessment of intake

Pesticide residue intake through the diet can be predicted with different degrees of accuracy. However, the more realistic predictions involve the consideration of many factors and therefore may be rather time-consuming. The options in the process are shown in Fig. 1.

Fig. 1. Options for the prediction of dietary intake of pesticide residues

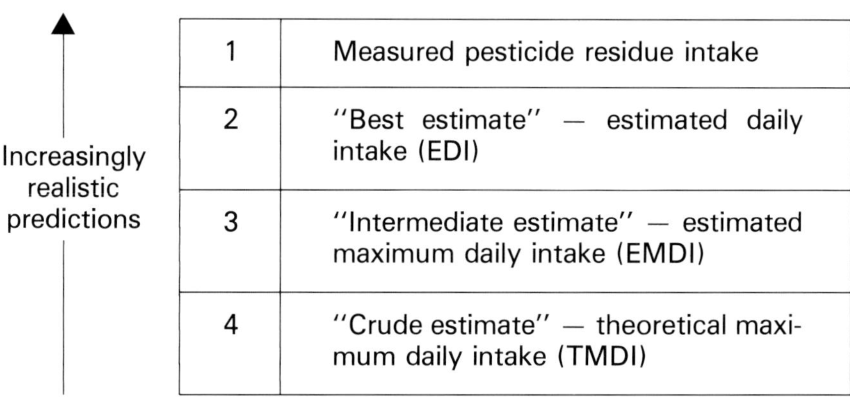

The procedures described here start with the most exaggerated and proceed towards more and more realistic intake predictions. It should be noted that the less realistic predictions, which are relatively straightforward to carry out, give an overestimate of the true pesticide intake. By starting with the most exaggerated predictions, it is therefore possible to eliminate at an early stage pesticides whose intake is clearly unlikely to exceed the ADI. More realistic predictions using refined data then make it possible to eliminate other pesticides from further consideration. Such an approach would facilitate acceptance of Codex MRLs for the majority of pesticides and allow the national authority concerned to direct its attention to those most likely to be of public health concern. The three-tier approach to predicting pesticide residue intake outlined in Fig. 2 is therefore proposed.

Fig. 2. Outline of proposed procedures for predicting pesticide residue intake

	TMDI[a]	EMDI[a]	EDI[b]
Residue level	Codex or national MRL.	Codex or national MRL. Corrections for: (i) edible portion; and (ii) losses on storage, processing, and cooking.	Known residue level. Corrections for: (i) edible portion; and (ii) losses on storage, processing, and cooking.
Food consumption	Hypothetical global or national diet. All commodities with a Codex or national MRL.	"Cultural" or national diet. All commodities with a Codex or national MRL.	National diet. Known uses of pesticide, taking account of: (i) range of commodities; (ii) proportion of crop treated; and (iii) home-grown and imported crops.

[a] May be estimated at either the national or international level.
[b] Can be estimated only at the national level.

Theoretical maximum daily intake (TMDI)

The TMDI is an estimate of dietary intake calculated using the MRL and the average daily per capita consumption of each food commodity for which an MRL has been established. The TMDI is calculated by multiplying the MRL by the average food consumption for each commodity and then summing the products:

$$TMDI = \sum F_i \times M_i$$

where

F_i = the average food consumption for the relevant commodity, as derived from the hypothetical global or national diet in kg of food per person per day; and

M_i = the MRL for the relevant commodity in mg of pesticide per kg of food.

Thus, the TMDI is given in units of mg per person. The ADI is, however, expressed in units of mg of pesticide per kg of body weight. In order to compare the TMDI with the ADI, the TMDI is divided by an assumed average body weight (this is usually taken to be 60 kg).

The TMDI will be a gross overestimate of the true pesticide residue intake because:

- the proportion of a crop treated with a pesticide is usually far less than 100%;
- very few of the crops treated with a pesticide contain the maximum residue level;
- residues are normally dissipated during storage, transport, preparation, commercial processing, and cooking of the treated commodity; and
- the MRL applies to the whole raw agricultural commodity, which frequently includes inedible portions. A large proportion of the residue may thus be discarded upon removal of the inedible portion.

It should therefore not be concluded that proposed Codex MRLs are unacceptable when the TMDI exceeds the ADI. Instead, a TMDI calculation should be used only as a screening process that may eliminate the need for further consideration of the intake of a pesticide residue.

On the other hand, if the TMDI does not exceed the ADI, it is highly unlikely that the ADI would be exceeded in practice, provided that the main uses of the pesticide are covered by the Codex MRL. Thus, more refined predictions of pesticide residue intake are not necessary.

Estimated maximum daily intake (EMDI)

The EMDI is a more realistic prediction of the pesticide residue intake. It is calculated using data on the edible portion of the commodity and takes into account the effects of the preparation, processing, and cooking of food, as follows:

$$EMDI = \sum F_i \times R_i \times P_i \times C_i$$

where

F_i = food consumption for the relevant commodity as derived from a specific hypothetical "cultural" or national diet in kg of food per person per day;

R_i = the residue level in the edible portion of the commodity given in mg of pesticide per kg of food (see Annex 2, Note A);

P_i = a correction factor that takes into account the reduction or increase in the residue on commercial processing, such as canning or milling (see Annex 2, Note B);[a] and

C_i = a correction factor that takes into account the reduction or increase in the level of residue on preparation or cooking of the food (see Annex 2, Note C).[a]

The units of the EMDI are the same as for the TMDI (mg of pesticide per person). In order to compare the EMDI with the ADI, the EMDI is therefore divided by an assumed average body weight (usually 60 kg), as in the comparison of the TMDI with the ADI.

Although the EMDI is a more realistic estimate of true pesticide residue intake than the TMDI, it is still an overestimate because it does not take into account that:
- the proportion of a crop treated with a pesticide is usually far less than 100%; and
- very few of the crops treated contain residue levels as high as the MRL, from which R is usually derived.

If the EMDI exceeds the ADI, it will be necessary to try to estimate more closely the true intake, as described below.

Estimated daily intake (EDI)

Calculation of the EDI takes into account the following factors:
- data on food consumption, including that of subgroups of the population;
- known uses of the pesticide concerned (see Annex 2, Note D);
- known residue levels (see Annex 2, Note E);
- the proportion of the crop treated;
- the ratio of the amount of home-grown to imported food; and
- the reduction in the level of pesticide during storage, processing, and cooking.

Since this type of information is usually only available at the national level, EDI predictions can only be performed on a national basis by those who have adequate information on food consumption, the use of a given pesticide locally, and the nature and the amount of imported food.

---

[a] The correction factors for residue losses during processing or cooking may be derived from information given in various JMPR evaluations (see, for example, 6).

3. USE OF THE GUIDELINES

Fig. 3 depicts how predictions could be used to assess the safety of pesticide residues by comparison with the ADI and to gauge the acceptability of Codex MRLs.

Fig. 3. Schematic representation of the relationships between relevant factors used in the guidelines

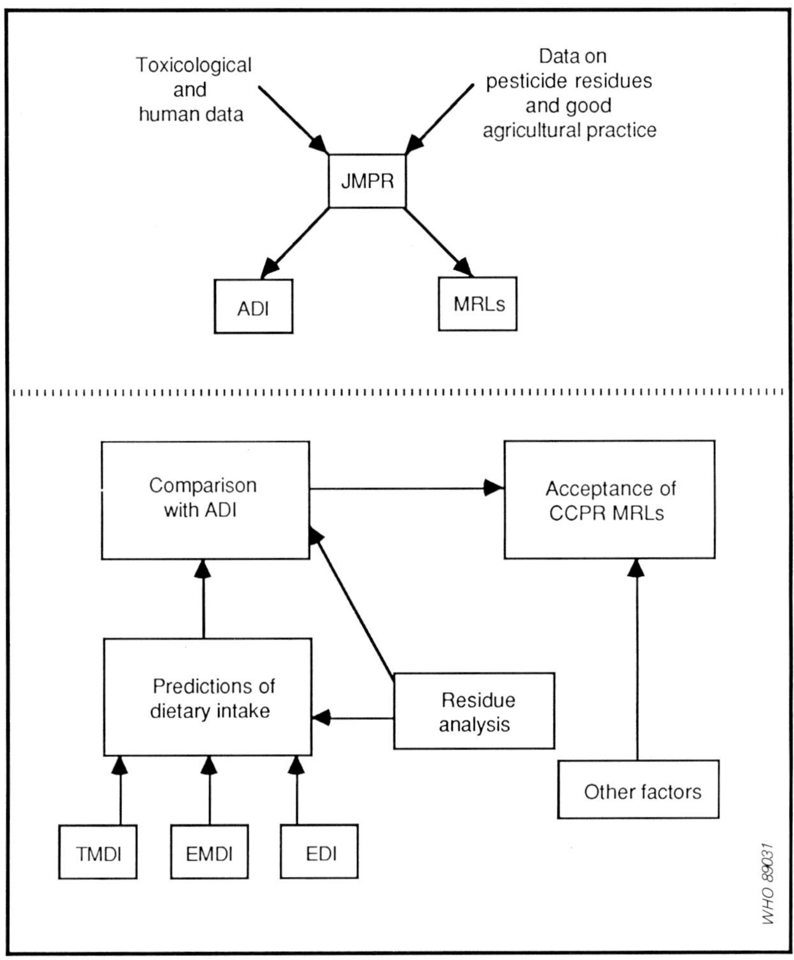

It is desirable that any presentation of the estimates be kept as simple as possible.

The need to guard against misuse of such calculations is strongly emphasized and under no circumstances should estimates be used to attempt to deduce the true pesticide residue intake, and hence to measure the level of consumer exposure (5).

As an illustration of the use of these guidelines, the TMDI and EMDI values for a hypothetical pesticide have been calculated. For this example, an arbitrary diet was selected and the MRLs used for the listed commodities are those applicable to several pesticides in common use. It can be seen that, while the TMDI gives a figure that is 75% greater than the fictional ADI (Table 1), the inclusion of more realistic information in the EMDI calculation brings this figure down to 25% of the ADI value (Table 2). In this case, an EDI calculation is unlikely to be required. However, in view of the high proportion of the EMDI that is due to residues on rice, authorities in areas where this commodity represents a greater proportion of the diet than in this example may wish to estimate also the EDI.

The correction factors for residue losses on processing or cooking have been derived from information given in various JMPR evaluations in which the results of appropriate studies were reported for several pesticides.

Table 1. Calculation of TMDI (ADI for pesticide X = 0.02 mg/kg of body weight)

Commodity	Food consumption (kg per person per day)	MRL (mg/kg)	TMDI (mg/person)
Wheat	0.11	5	0.55
Rice	0.22	5	1.10
Apples	0.04	2	0.08
Bananas	0.08	1	0.08
Citrus fruits	0.03	5	0.15
Cabbage	0.10	0.5	0.05
Lettuce	0.02	0.5	0.01
Potatoes	0.40	0.2	0.08
Cattle meat	0.20	0.05[a]	—
Milk	0.30	0.01[a]	—
Total			2.10 (0.035 mg/kg body weight)[b]

[a] At or about the limit of determination.
[b] Equivalent to 175% of the hypothetical ADI.

Table 2. Calculation of EMDI (ADI for pesticide X = 0.02 mg/kg of body weight).

Commodity	Processed commodity	Food consumption (kg per person per day)	Residue level (mg/kg)	Processing factor	Cooking factor	EMDI (mg/person)
Wheat	Bread	0.11	5	0.16	0.038	0.003
Rice	—	0.22	5	1	0.14	0.154
Apples	—	0.04	2	1	1	0.080
Bananas	Edible flesh	0.08	0.05	1	1	0.004
Citrus fruits	Edible flesh	0.03	0.1	1	1	0.003
Cabbage	—	0.10	0.5	1	0.5	0.025
Lettuce	—	0.02	0.5	1	1	0.01
Potatoes	—	0.40	0.2	1	0.5	0.04
Cattle meat	—	0.20	0.05[a]	—	—	—
Milk	—	0.30	0.01[a]	—	—	—
Total						0.319 (0.005 mg/kg body weight)[b]

[a] At or about the limit of determination.

[b] Equivalent to 25% of the hypothetical ADI.

REFERENCES

1. UNEP/FAO/WHO. *Guidelines for the study of dietary intakes of chemical contaminants.* Geneva, World Health Organization, 1985 (WHO Offset Publication No. 87).

2. FAO. *Pesticide residues in food – 1981. Report of the Joint Meeting of the FAO Panel of Experts on Pesticide Residues in Food and the Environment and the WHO Expert Group on Pesticide Residues, Geneva, 23 November – 2 December 1981.* Rome, Food and Agriculture Organization of the United Nations, 1981 (FAO Plant Production and Protection Paper 37).

3. *Codex limits for pesticide residues in food and consumer safety.* Discussion paper by the *Ad Hoc* Working Group on Regulatory Principles, Codex Committee on Pesticide Residues (Unpublished document CX/PR 86/12, February 1986).

4. FAO. *Food balance sheets: 1979-81 average.* Rome, Food and Agriculture Organization of the United Nations, 1984.

5. FAO. *Pesticide residues in food – 1986. Report of the Joint Meeting of the FAO Panel of Experts on Pesticide Residues in Food and the Environment and the WHO Expert Group on Pesticide Residues, Rome, 29 September – 8 October 1986.* Rome, Food and Agriculture Organization of the United Nations, 1986 (FAO Plant Production and Protection Paper 77).

6. FAO. *Pesticide residues in food – 1987. Report of the Joint Meeting of the FAO Panel of Experts on Pesticide Residues in Food and the Environment and the WHO Expert Group on Pesticide Residues, Geneva, 21-30 September 1987.* Rome, Food and Agriculture Organization of the United Nations, 1988 (FAO Plant Production and Protection Papers 84, 86/1, and 86/2).

GLOSSARY [a]

Acceptable daily intake (ADI)

The ADI of a chemical is the daily intake which, during a lifetime, appears to be without appreciable risk, on the basis of all the facts known at the time. It is expressed in milligrams of the chemical per kilogram of body weight.

Codex Committee on Pesticide Residues (CCPR)

The CCPR is a subsidiary body established by the Codex Alimentarius Commission. The CCPR has the responsibility for establishing maximum residue limits for pesticides in food and feed, to prepare priority lists of pesticides for evaluation by the JMPR (see p. 19), to consider methods of sampling and analysis for the determination of pesticide residues in food and feed, and to consider other matters in relation to the safety of food and feed that contain pesticide residues. Membership of the CCPR is open to all Member States and Associate Members of FAO and WHO. Representatives of international organizations that have formal relations with either FAO or WHO may attend meetings as observers. The CCPR is hosted by the Government of the Netherlands and has met twenty times since 1966.

Codex maximum residue limit (MRL)

A Codex MRL is defined as the maximum concentration of a pesticide residue, resulting from the use of a pesticide according to good agricultural practice, that is recognized by the Codex Alimentarius Commission to be legally permitted or acceptable in or on a food, agricultural commodity, or animal feed. The MRL is expressed in milligrams of the residue per kilogram of the commodity.

"Cultural" diet

In the context of this publication, a "cultural" diet is a hypothetical diet representative of dietary patterns in which the quantitative intake of food groups is similar.

---

[a] The definitions given in this glossary are for use with this publication only, and are not necessarily of general validity.

Estimated daily intake (EDI)

The EDI is a prediction of the daily intake of a pesticide residue
based on the most realistic estimation of residue levels in food and
the best available food consumption data for a specific population.
The residue levels are estimated taking into account known uses of
a pesticide, the range of contaminated commodities, the proportion
of a commodity treated, and the quantity of contaminated home-
grown or imported commodities. The EDI is expressed in
milligrams of the residue per person.

Estimated maximum daily intake (EMDI)

The EMDI is a prediction of the maximum daily intake of a
pesticide residue based on the assumptions of average daily food
consumption per person and maximum residues in the edible
portion of a commodity, corrected for the reduction or increase in
residues resulting from preparation, cooking, or commercial processing. The EMDI is expressed in milligrams of the residue per
person.

Food consumption

Food consumption is an estimate of the daily average per capita
quantity of a food or group of foods consumed by a specified
population. Food consumption is expressed in kilograms of food
per person per day.

Good agricultural practice

Good agricultural practice in the use of pesticides is the officially
recommended or authorized use of such substances, under practical
conditions, at any stage of production, storage, transport, distribution, or processing of food, agricultural commodities, or animal
feed, bearing in mind the variations in requirements within and
between regions. This takes into account the minimum quantities
necessary to achieve adequate control, applied in such a manner
that the amount of residue is the smallest practicable and which is
toxicologically acceptable.

Joint FAO/WHO Meeting on Pesticide Residues (JMPR)

The JMPR is the abbreviation for the Joint Meeting of the
FAO Panel of Experts on Pesticide Residues in Food and the
Environment and the WHO Expert Group on Pesticide Residues.

Such meetings are normally convened annually, and during them the FAO Panel of Experts is responsible for reviewing pesticide use patterns (good agricultural practice), data on the chemistry and composition of pesticides, methods of analysing pesticide residues, and for estimating the maximum residue levels that might occur following the use of a pesticide according to good agricultural practice. The WHO Expert Group is responsible for reviewing toxicological and related data on the pesticides and, where possible, for estimating ADIs for humans.

No-observed-adverse-effect level (NOAEL)

The NOAEL is the highest dose of a substance in experimental animal studies that does not cause any detectable toxic effects. The NOAEL is expressed in milligrams of the substance per kilogram of body weight per day.

Pesticide residue

A pesticide residue is any specified substance in food, agricultural commodities, or animal feed resulting from the use of a pesticide. The term includes any derivatives of a pesticide, such as conversion products, metabolites, reaction products, and impurities that are considered to be of toxicological significance.

Risk

Risk is a statistical concept defined as the expected frequency of undesirable effects arising from exposure to a chemical. It may be expressed as absolute risk (excess risk due to exposure) or relative risk (the ratio of the risks in exposed and unexposed populations).

Theoretical maximum daily intake (TMDI)

The TMDI is a prediction of the maximum daily intake of a pesticide residue, based on the assumptions of MRL levels of residues in food and average daily food consumption per person. The TMDI is expressed in milligrams of residue per person.

JOINT FAO/WHO CONSULTATION ON GUIDELINES FOR PREDICTING DIETARY INTAKE OF PESTICIDE RESIDUES

Geneva, 5-8 October 1987

Participants

Dr D.C. Abbott, London, England.

Mr J.A.R. Bates, Harpenden, England.

Dr A.L. Black, Health Research and Services Division, Department of Community Services and Health, Woden, ACT, Australia (*Chairman*).

Mr D.J. Clegg, Toxicological Evaluation Division, Food Directorate, Health Protection Branch, Health and Welfare Canada, Ottawa, Ontario, Canada (*Rapporteur*).

Mr J. van der Kolk, Ministry of Welfare, Health and Culture, Rijswijk, Netherlands (*Vice-Chairman*).

Representatives of other organizations

Mr A.J. Pieters, Chairman, Codex Committee on Pesticide Residues, Ministry of Welfare, Health and Culture, Rijswijk, Netherlands.

Mr G.A. Willis, Chairman, GIFAP Residue Committee, ICI Plant Protection Division, Haslemere, Surrey, England.

Observer

Dr C.E. Fisher, Food Safety and Surveillance Unit, Food Science Division, Ministry of Agriculture, Fisheries and Food, London, England.

Secretariat

Dr G. Gheorghiev, Food Quality and Standards Service, Food Policy and Nutrition Division, Food and Agriculture Organization of the United Nations, Rome, Italy (*Joint Secretary*).

Dr H. Galal Gorchev, Food Safety, Division of Environmental Health, World Health Organization, Geneva, Switzerland (*Joint Secretary*).

Dr J. Herrman, International Programme on Chemical Safety, Division of Environmental Health, World Health Organization, Geneva, Switzerland.

Dr F.K. Käferstein, Food Safety, Division of Environmental Health, World Health Organization, Geneva, Switzerland.

Dr R.D. Schmitt, Office of Pesticide Programs, Environmental Protection Agency, Washington, DC, USA (*Rapporteur*).

Dr G. Vettorazzi, International Programme on Chemical Safety, Division of Environmental Health, World Health Organization, Geneva, Switzerland.

NOTES

A. Portions of the commodity consumed

Residues of non-systemic pesticides that occur on the surface of melons, squashes, cantaloups, bananas, avocados, kiwi fruit, pineapples, and similar crops are not consumed, since the peel is discarded. Thus, residue levels in the edible parts of these fruits should be used in place of the maximum residue levels in the whole fruit. Analysing both the whole commodity and the edible portion increases the cost of analyses and thus often limits the generation of data on the edible portion. However, data on residue levels in the edible portion are essential for the realistic prediction of the pesticide residue intake.

These considerations also apply to citrus fruits, but intake of residues from processed commodities, such as orange juice, orange oil, and ground peel should not be ignored. However, consumption of some processed commodities is relatively low compared with that of the whole fruit and juice, and the pesticide residue intake from such sources can usually be ignored.

B. Effects of commercial processing on levels of pesticide residues

Many commodities are processed before consumption. For example, cereal grains are major foods in most countries and are usually consumed after milling. Residue levels in some milling fractions, such as white flour, are almost always lower than those in the whole grain. On the other hand, residues in certain milling fractions, such as bran, may contain higher levels of residue than the whole grain. Lipid-soluble pesticides that concentrate in crude vegetable oils are frequently removed by the refining processes used to make the oil suitable for human consumption. When data are available on the residue levels in processed commodities, the use of residue levels in the processed commodity, instead of the maximum level in the whole commodity, leads to a more realistic prediction of pesticide intake than the TMDI.

C. Effects of preparation and cooking on levels of pesticide residues in food

The preparation of many commodities for consumption involves operations that result in a reduction in pesticide residues. Washing food removes some, and occasionally most, of a surface residue. Removing the outer leaves from a crop food will frequently result in considerably lower residues of non-systemic pesticides than those that occur in the commodity as it moves in commerce.

Many pesticides are thermally unstable and are also hydrolysed in the presence of water. For such pesticides, residue levels in cooked commodities are often well below the MRL. Cooking may be a part of commercial canning operations or may be carried out in a restaurant or in the home. The cooking process will depend on the commodity, but the most common forms are baking, boiling in water, frying in oil, and grilling. Many fruits and vegetables can be eaten either raw or cooked, and the assumption that all vegetables are cooked can lead to an underestimation of the true intake of pesticide residues.

D. Known uses of a pesticide

In estimating both the TMDI and EMDI at the international level, it is assumed that pesticide residues are present only in commodities for which there are Codex MRLs. The prediction of the EDI, usually carried out at the national level, requires information on the known uses of the pesticide on both home-grown and imported foods. This may include commodities for which there are no Codex MRLs, but may also exclude commodities for which there are Codex MRLs, but for which the country concerned knows that there is no use of the pesticide on either home-grown or imported foods.

E. Known pesticide residues

For a better prediction of the EDI value at the national level, it is necessary to have some information about the level of the pesticide residue that is most likely to occur in the commodity in practice; this may well vary from country to country, for many reasons. Such information can be derived from various sources including supervised trials, survey sampling and analysis, monitoring data, and the mode and time of application of the pesticide.